HABITATS

ISLANDS

JULIA WATERLOW

Wayland

HABITATS

Coasts
Deserts
Forests
Grasslands
Islands
Mountains
Polar Regions
Rivers and Lakes
Seas and Oceans
Wetlands

Cover: An aerial view of islands in a warm tropical sea.
Contents page: The Komodo Dragon of Indonesia.

Series and book editor: Rosemary Ashley
Series designer: Malcolm Walker

First published in 1995 by
Wayland (Publishers) Limited
61 Western Road, Hove
East Sussex, BN3 1JD, England

British Library Cataloguing in Publication Data
Waterlow, Julia
Islands. - (Habitats series)
I. Title II. Series
910.9142

ISBN 0-7502-1487-2

Typeset by Kudos Editorial and Design Services
Printed and bound in Italy by L.E.G.O. S.p.A., Vicenza, Italy

CONTENTS

1. SURROUNDED BY WATER

Far from land, hidden among the vast watery expanses of the oceans, islands conjure up dreams of mystery, adventure and castaways. For some of us they are a kind of paradise, covered with palm trees, fringed by sandy beaches and surrounded by clear warm blue seas. In fact there are still thousands of small islands in far-off places, uninhabited and little visited by people, that fit this romantic ideal. However, in the modern world many islands have become important and busy centres of trade, heavily built-up and densely populated.

The general name given to the place where plants or creatures live is called their habitat. An island differs from other habitats because it is surrounded on all sides by water. If the island is a long way from the mainland, the water can act like a barrier. As a result, some islands support their own special creatures and plants that may not be found anywhere else. The fact that islands may be isolated from the rest of the world can affect the way the human island population has developed too.

This map indicates some of the world's larger islands and island groups.

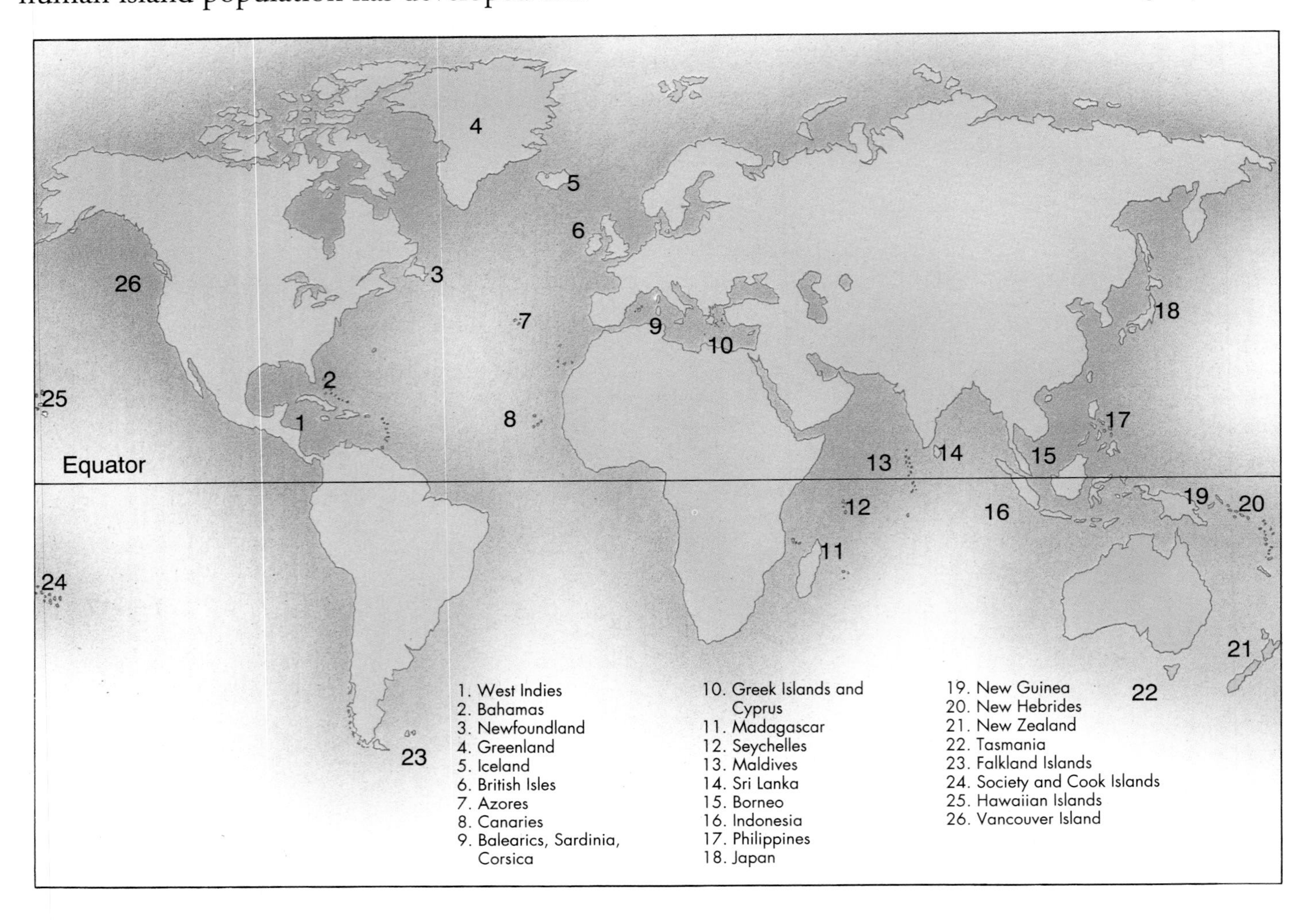

Left Sun, sand and palm trees, surrounded by clear water – perhaps the ideal desert island.

Although islands have in common the fact that they have water all around, the habitats they provide are very varied. There is every type of landscape, from ice-covered lumps of rock near the North and South Poles, to the lush tropical forests of islands close to the equator. In many cases it is not just the climate but the way the island was formed that allows certain types of plants and creatures to survive.

History, politics and economics have shaped the way islands are today just as much as their physical characteristics. Outsiders have appeared, conquering, invading or trading, and have changed the way of life for native islanders for ever. On many islands new plants and animals have been introduced, permanently altering the islands native species. As in many parts of the world, the natural resources of islands are being heavily exploited. Care of the environment is of special concern because most islands have fragile ecologies and the natural balance can be easily upset.

Below The heart of New York, in the USA, is crammed on to Manhatten Island.

2. ISLAND FORMS

Although all islands are surrounded by water, it is not always the sea; islands are also found in lakes and rivers. Some islands have existed for millions of years. Others only appear for a short space of time, perhaps a few hundred years or even less, quickly eroded away by the forces of nature, disappearing under the sea if the water-level rises or if the earth's crust sinks.

There are four main ways that islands can be created. Land sometimes becomes separated from the mainland, or a mound of earth is built up as mud or sand is deposited by water. Other islands are formed by volcanic action, or by the growth of coral.

Continenal islands

Many islands, often those lying near a continent, were once joined to the mainland. About 220 million years ago, the continents as we know them today were once part of one huge landmass called Pangaea. This gradually broke up, and the pieces, or continents, drifted apart, eventually reaching their present positions and shapes. As this happened, smaller pieces separated off from the larger continents and became islands surrounded by sea. The great island of Madagascar, off the south-east coast of Africa, was formed in this way. About 160 million years ago the island broke away from the African continent and has been separated ever since by a 400 kilometre strait of water.

Sea level changes have also left pieces of land that were once part of a continent separated by water. One reason the sea-level varies in height is that

A diagram showing the break up of continents.

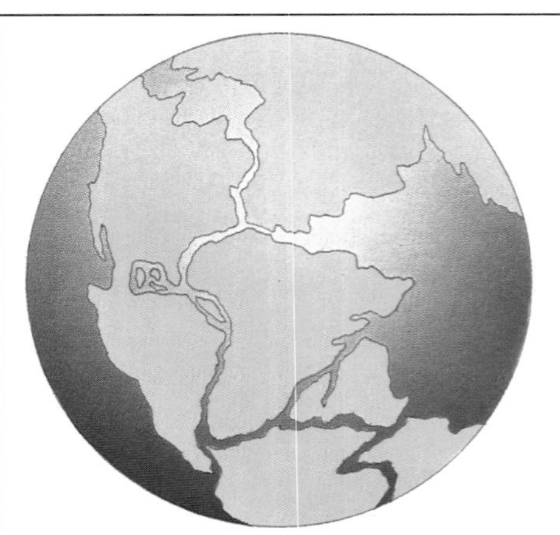

200 million years ago. At this time there was only one huge landmass surrounded by ocean. India was wedged between Africa and Australia.

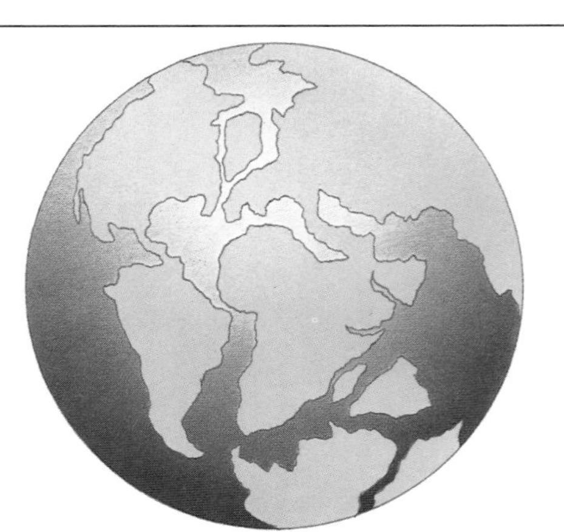

65 million years ago. As the huge landmass gradually broke up, pieces of it drifted apart. Oceans were formed, separating the continents. Very gradually, over millions of years, the continents moved position.

Today. The continents are in their present, familiar position. India has drifted to join Europe and Asia, and Australia has drifted far away from the original landmass.

the climate of the earth is constantly changing. The last cold period, the last Ice Age, peaked about 18,000 years ago. Freezing temperatures turned the seas to ice and massive sheets of ice several kilometres thick moved out from the poles. So much water was locked up as ice that it is thought the level of the sea was more than 120 m lower than it is now. When the earth warmed again the ice melted and raised the level of water in the sea all over the world.

Britain and most of its many islands were once linked by a land bridge with the European continent. During the last 8,000 years, after the end of the last Ice Age, the sea-level gradually rose, flooding valleys and lowland areas until the British Isles became separated completely from Europe by the English Channel. Sri Lanka, Tasmania, Taiwan and many of North America's east coast islands, such as Manhattan and Newfoundland, were also once part of the mainland they adjoin.

Sedimentary islands

Islands can appear where mud or soil is deposited in great quantities. Off the coasts of The Netherlands, Germany and Denmark lie the Friesan Islands, a chain of low sandy islands. Sand and soil were gradually dumped along this part of the coast by the sea, forming sand dunes. These were slowly broken up and separated from the mainland by storms. Today there is a shallow sea between the islands and the mainland. They support good grazing for cattle but are too sandy and windswept for trees to grow.

One of the Friesan Islands off the coast of Germany. These islands were formed by soil and sand deposited over thousands of years.

Islands often appear near the mouths of rivers. These are usually made up of sediment, the mud and earth carried by a river. In its early stages a river is fast-flowing and powerful and erodes the surrounding earth and rock, carrying it away downstream with the water. A river slows near the end of its course, and as it does so, it begins to drop its load of sediment on the river-bed. In some places so much sediment is deposited that it forms a mound of mud and soil that grows into an island.

In the mouth of the Amazon River (in Brazil) lies Marajo Island, the largest river delta island in the world. It is nearly the size of Switzerland. Large parts of the island are regularly flooded in the wet season although enough remains permanently dry for people to live on the island and trees to grow.

Volcanic islands

Thousands of the world's islands were and still are being created by volcanic activity. Nearly all islands lying in the middle of oceans appeared in this way.

Underneath the crust of our planet it is so hot that the rocks and minerals are a molten liquid. On top of this, like a skin on custard, the earth's surface is made up of a series of solid plates that float and slowly move around. Where the plates meet and rub together are areas where the crust is weakest and the molten interior of the earth can spill out, sometimes in a volcanic explosion or at other times in a seeping flow of lava. Many of the earth's plates meet under the sea. Often the volcanic activity at these points throws up ash and lava that solidifies when it cools on contact with the air or water. Where the eruption is big enough, an island emerges out of the sea.

Most of the world's volcanic islands appear in the Pacific Ocean, where the vast Pacific plate collides with other larger and smaller plates. One of the

An appearing and disappearing island

In the eighteenth century Falcon island, near Tonga, in the Pacific Ocean, was recorded as a reef. In 1885 the island rose 80 m above sea-level, but by 1899 it had disappeared again because of erosion. In 1927 a volcanic eruption raised the island again, to a height of 100 m above sea-level but by 1949 it had sunk once more.

Grey lava fills a crater on Kilauea volcano, Hawaii, an island in the middle of the Pacific Ocean.

best-known Pacific islands is Hawaii. It is one of a line of volcanic islands, the first of which, Kauai, appeared about five-and-a-half million years ago. Hawaii is the youngest in the chain, appearing about one million years ago. It has three volcanoes, the highest rising to 4,200 m above sea-level. Kilauea volcano still oozes molten lava – the last major flow occurred in 1990.

Volcanic islands can appear and disappear overnight or take hundreds of years to appear as the lava slowly builds up. One of the newest and quickest island arrivals was the island of Surtsey, in the North Atlantic not far from Iceland. Surtsey lies along another important oceanic line of volcanic activity, the Mid-Atlantic Ridge. In November 1963 lava started bulging up out of the sea. Suddenly an eight-kilometre-high column of steam, ash and fumes poured out; at some moments rocks were being thrown out from the volcano at a rate of half a million tonnes per hour. In two days the island was 40 m high; within a week it had reached 100 m high and nearly a kilometre long. The eruptions finally stopped in 1967, leaving an island about three kilometres in length.

Spectators on Surtsey watch a nearby volcanic island erupting out of the sea.

Island explosion

On 28 August 1883 hot molten rock and ash was blasted into the stratosphere, some of it 80 km into the sky. In a few seconds the volcanic island of Krakatoa disappeared as it collapsed, causing the biggest natural explosion ever recorded. The deafening noise even reached Australia, 5,000 km away. The explosion caused a huge tidal wave and a wall of water the height of a four-storey house, travelling at 500 kph. It wrecked the shores of the island of Java nearby; in South Africa, 8,000 km away, boats were rocked at anchor.

A coral island surrounded by reefs, in the Mariana Islands in the Pacific Ocean.

Coral islands

There is one type of island that is actually made by a living creature: the coral island. Millions of tiny animals build tough shell-like homes, each on top of another to form a reef, a kind of hard underwater bank, in the sea. Should the sea-bed rise or sea-level fall even slightly, these reefs are left above the sea as islands.

Coral likes shallow, clear and warm water and is often found growing around volcanic islands in tropical seas. Many islands of the Pacific and

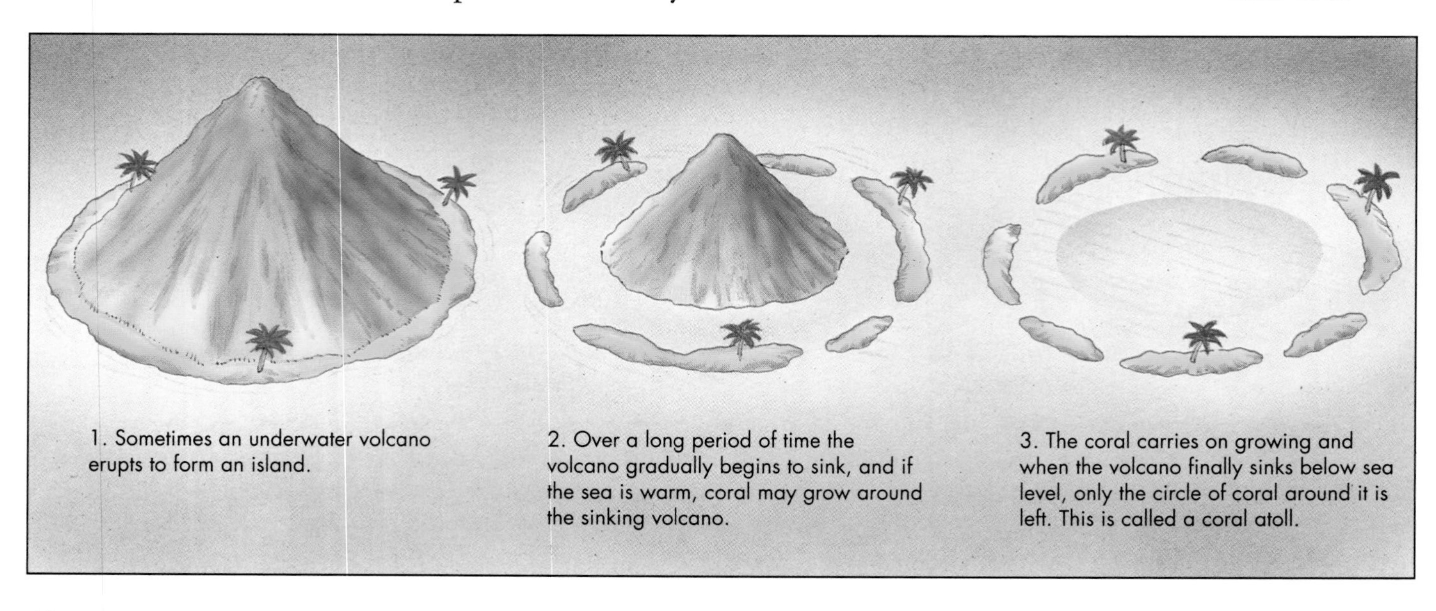

A diagram showing the formation of a coral atoll.

Indian Oceans are made of coral. Coral grows incredibly slowly – even in the warmest waters it only grows two to three metres in a century. It can happen that a volcanic island slowly drops back into the ocean and only the coral ring that has grown around the volcano is left, making a circle of coral reefs. This is called an atoll and can be anything from a few kilometres across to several tens of kilometres across. The reef is seldom continuous and has breaks in it, channels that allow boats to sail through into the inner lagoon.

Aldabra is an atoll in the Seychelles, lying 400 km east of Africa. It is only 30 km long and easy to miss in the seas of the great Indian Ocean. Its huge lagoon is encircled by a ring of coral islets that sit like a cap on an underwater volcano rising 4,000 m from the sea-bed below. Aldabra is known to have appeared above the water and dropped below its surface several times as the sea-level or ocean floor have risen or fallen; its rise to its present height is thought to have begun about 50,000 years ago.

Land or sea-level changes have left this coral exposed on Nuaru Island, in the Pacific.

Facts and figures

- *Although Australia is completely surrounded by sea, it is usually called a continent and not an island because of its huge size.*
- *The largest island in the world is Greenland. It is 1,176,000 sq km in area.*
- *The other nine largest islands in descending order of size are: New Guinea, Borneo, Madagascar, Baffin Island (Canada), Sumatra, Honshu (Japan), Britain, Vancouver Island (Canada), Ellesmere Island (Canada).*
- *The largest freshwater island is Marajo Island in the mouth of the Amazon River. It is about 250 km wide by 200 km long.*
- *The largest atoll in the world is Kwajalein in the Marshall Islands (Pacific Ocean). It is 283 km long, enclosing a lagoon 2,850 sq km in size.*
- *The world's most remote island is Bouvet Oya in the South Atlantic. It lies 1,700 km from the nearest land.*
- *The highest island mountain is Hawaii's Mauna Kea that soars up to a height of 4,205 m.*

3. LIFE SUPPORT

The type and variety of natural life an island can support depends on the climate, the kind of soils that exist and what seeds or creatures manage to reach the island to start life there.

The climate of an island varies with its latitude; the Arctic and Antarctic islands all suffer from freezing temperatures for much of the year and this encourages little more than lichens and mosses to grow. As well as being the largest island in the world, most of Greenland lies within the Arctic Circle and so it is one of the coldest islands. Much of Greenland is covered by ice and snow all year round. Islands nearer the equator are warmer and so usually are able to support more life. Borneo, in Southeast Asia, lies directly on the equator and has dense tropical forests with a huge variety of plants and animals.

Islands do have one climatic feature in common: because they are surrounded by sea or large bodies of water, this has the effect of keeping temperatures moderate (even). Water retains heat longer than land so in colder months places near the sea tend to be warmer than those inland. The opposite is true in the summer – coastal areas are cooled by the sea.

The bitterly cold island of Greenland supports little natural vegetation.

Being surrounded by water can also mean a comparatively high rainfall. Britain is one example. From the study of pollen in ancient soils it is thought that Britain's climate changed when the country became an island some 8,000 years ago. Ocean water started circulating round the land and the weather changed from a drier continental climate to a wetter one, kept moderate by the Atlantic Ocean. Evidence for this comes from the way pine, a continental tree, was replaced by the water-loving alder tree.

Above Lush greenery covers the warm, wet, tropical island of Bali, in Indonesia.

Many islands suffer from too much rain falling at once, especially those in tropical areas such as the Caribbean and the Philippines. These regions are particularly prone to cyclones, hurricanes and typhoons (all different names for tropical storms), which can be very destructive. When in 1989 Hurricane Hugo struck the Caribbean island of Montserrat, it left 95 per cent of the island's buildings damaged, roads washed away and plants uprooted.

Plants not only need warmth and water to grow but also soil. A brand new island such as a recently formed volcanic or coral island starts off with bare rock. However, very quickly the wind, rain and sun get to work in a process

Left Hurricane damage on Caribbean islands often leaves a scene of devastation.

The bleak volcanic landscape of Ascension Island, in the Atlantic Ocean, where little plant life has been able to take root.

called weathering. Gradually tiny bits of rock break up and collect in crevices. This happens all over the rocky island until there are little pockets of soil in which seeds find a foothold and grow.

Few islands remain deserts for long. Soon, from across the sea, come seeds and small creatures. The wind can carry light plant seeds long distances from the mainland or a nearby island; birds bring seeds either in their droppings or perhaps stuck to their feet; floating driftwood that reaches the island may carry small creatures such as ants, spiders or beetles. Often grass and other loose plants mat together, making a sea-going platform on which larger animals may sometimes travel. Reptiles tend to survive particularly well on such

Because they can swim, turtles are often the first creatures to inhabit islands.

Life after Death

A hundred years after the massive explosion of Krakatoa (see page 9), a small part of the old island still remains. Slopes that a century ago were bare volcanic rock are now covered by thick tropical forest and are full of wildlife. Seeds floated across the sea or were brought by birds. Winged creatures such as birds, butterflies and insects flew from the mainland. The rats, pythons and lizards on the island probably arrived on logs or other floating objects.

The first plants appear on Surtsey.

journeys as they have watertight skin which is not much harmed by sea-water. Creatures that can swim, such as turtles, are some of the first to reach and inhabit an island. If humans land there, they can unsuspectingly bring other land creatures, such as rats that have been living in the ship.

There is no knowing how fast plant life will appear and what species will take hold. Much depends on the direction of winds, how close other land lies and what plants and creatures manage to make the journey across the water. Over time, the variety and numbers will grow. On the distant but comparatively old island of Hawaii, far out in the Pacific Ocean, over 1,700 flowering plants alone managed to reach the island by natural means. The relatively young Maldive Islands have only 87 different naturally introduced plants and about twice that number of species of animals.

The coconut

The coconut is a typical tropical island plant as it has found a way to travel across the world by water. The coconut palm grows along the edge of tropical beaches, leaning over so that when the coconuts fall they roll into the sea. The thick husk of the coconut allows it to float and protects the nut inside for as long as four months at sea, until it lands on a new tropical island.

Below The different ways in which plant and animal life arrives on islands.

4. FLORA AND FAUNA

Because of the different kinds of habitats on islands around the world, the plants and animals vary from thin mosses to dense tropical forests, and from simple insects to the thousands of more complex species that thrive in the rich vegetation of the tropics. What is different about some islands is that because of their isolation, they have become sanctuaries for creatures and plants that are found nowhere else in the world.

Sanctuaries

One reason why isolated islands have become sanctuaries is because they have been naturally protected from changes that may have taken place elsewhere. Animals may still exist on an island because the predator (animal or human) or disease that killed them off in another place never reached the island. It might also be that there are fewer other creatures competing for the same food and so there is a better chance of survival on an island.

Madagascar has a number of creatures that are unique to the island. The most famous of these is the lemur, a long-tailed member of the monkey family. From fossils found on the African mainland scientists know that lemurs once lived there too but died out for an unknown reason. The separation of Madagascar from the mainland has allowed the lemur to evolve without the problems that affected it on the African continent.

Lemurs on Madagascar. They are found nowhere else in the world.

The islands of New Zealand are also home to creatures that became extinct elsewhere millions of years ago. The tuatara is a very rare creature that looks like a heavily-built lizard. It is actually related to dinosaurs and not to the modern lizard. The fossilized bones of an identical creature have been found in two-million-year-old rocks.

Another island that acts like a sanctuary is Aldabra in the Seychelles. One creature that lives there is the booby, a bird like a gannet. These birds fly for hundreds of miles searching for shoals of fish; they spend most of the year in the air. However, their chosen nesting site is Aldabra, one of the few uninhabited islands in the area that is virtually free of cats, rats and other creatures that might eat their eggs or chicks.

Above Nine types of these strange baobab trees grow on Madagascar. On the African mainland only one type grows.

Evolution

The isolation of islands not only protects plants and creatures from dying out but can also create a special environment where flora and fauna can evolve in a different way from anywhere else in the world. Because evolutionary changes are not weakened by large numbers of species as they are on the mainland, the speed of evolution is usually quicker on islands than elsewhere. Islands become small in-bred communities of creatures.

Left Flightless birds like this kiwi in New Zealand evolved on many islands that were free from predators.

When not basking on rocks, the iguanas of the Galapagos Islands spend much of their time in the sea.

The Galapagos Islands off the west coast of South America are home to several creatures that have evolved in a unique way. In particular are the thirteen species of finch found on the Galapagos and nowhere else. It is thought that originally (thousands of years ago) a few were blown here by a freak storm from the mainland. The islands have varied habitats, from desert areas to thick bush and high volcanic peaks, and the finches evolved and adapted to the different environments they found. Some finches have heavy beaks for crushing fruit, others have slim beaks to pick off insects, and one species even uses its beak to cut a cactus spine and push it into holes to get beetle grubs.

Other creatures found on the Galapagos are marine iguanas (the only lizards in the world that regularly spend time in the sea – they eat seaweed), huge tortoises and flightless cormorants. Flightless birds are common on many remote oceanic islands. Aldabra, for example, is home to the small rail bird which has similar cousins on mainland Africa. But the difference is that on Aldabra the rail cannot fly. On the mainland rails have many enemies and need wings to escape, but on Aldabra, where there are no animal predators, the bird has evolved without wings.

Probably the most famous of all flightless birds is the now-extinct dodo which used to live on the island of Mauritius in the Indian Ocean. It was a pigeon-like bird that grew to the size of a turkey and shed its wings as it evolved in isolation. New Zealand has its own version that has survived, the kiwi, the country's national symbol.

Giant tortoises are other creatures that have commonly evolved on remote islands. Madagascar and Aldabra are two of the places where they are found. The giant ones are probably descended from normal-sized tortoises which may have arrived on driftwood from the mainland. Tortoises have been found floating at sea, so it seems they can survive the long sea journey. It is not known exactly why they grow so large but it is possible that since there are no predators they just live longer and get bigger.

Growing to a large size is not restricted to tortoises. A hundred years ago there were tales of a monster on the islands of Indonesia that attacked humans. Known as the Komodo Dragon, the creature is in fact an overgrown monitor (a kind of lizard common in tropical countries). It grows up to 3 m in length, is much bulkier than a normal monitor, and eats meat. Because there are no other large meat-eating creatures to prey on pigs and deer on the island of Komodo, the lizard may have evolved in order to take advantage of this abundant source of food.

A giant Galapagos tortoise.

Galapagos

The Galapagos Islands lie on the equator, 970 km west of the coast of South America. The islands are the peaks of gigantic volcanoes, rising 2,000 to 3,000 m from the sea bed. There are six main islands, two of which are uninhabited, and over fifty smaller islands and islets. Many of the creatures of the islands are unique because they evolved in a different way from their ancestors on the South American continent. A quarter of the shore fish, half the plants and almost all the reptiles are found nowhere else. Since no large land mammals reached the islands, reptiles have become the dominant creatures, just as they were all over the world in the very distant past.

The Komodo Dragon, a giant lizard found on certain Indonesian islands.

5. ISLAND PEOPLES

Like the plants and creatures that inhabit islands, some island people have also developed their own particular characteristics. Being isolated and having little contact with others, they have languages, beliefs and traditions that are not found elsewhere. But over the last few hundred years, most islands have been visited by outsiders and, with modern transport and communications, few are still cut off from the world. This has meant changes for islanders as they adapt and learn about people elsewhere. Few are able to keep their traditional way of life as it once was.

Iceland is an island with one of the most non-mixed peoples in the world. It is a cold and isolated island in the North Atlantic that was settled by descendants of Nordic explorers and traders, the Vikings. There has been virtually no immigration to the island since it was first settled. In their isolation the Icelandic people developed their own culture, rich with sagas and poetry, and their language, related to Old English, is much as it was spoken originally.

On the other side of the world in the warmth of the Pacific Ocean, live the Polynesians. Polynesians have a light skin compared with other Pacific islanders such as Melanesians; they are tall and

Above Iceland, home to descendants of Viking explorers.

Left Polynesian girls from the Cook Islands, in the South Pacific.

well-built, mostly with straight dark hair and they all speak dialects of one language. This suggests that the Polynesians all came from one direction over a short period of time and had little chance to mix with other races. It remains a puzzle whether they originally came island-hopping eastwards from Asia or westwards across the vast ocean from the Americas.

Polynesians were and still are hardy seafarers, and they were culturally and technically advanced when they made their first journeys around the Pacific Islands. Where there was wood, they built strong houses and boats which they carved and decorated. They used wood and stone for tools, skins and the fibre from plants to make clothes, and farmed the land, using plants and fruits they took from island to island. As well as hunting and fishing, Polynesians domesticated animals like dogs, pigs and chickens. On many islands they built stone altars where they worshipped their ancestors. On some remote islands their traditional ways of life still remain largely untouched.

Also in the Pacific region lies the island of New Guinea. There are mountain tribes of the interior that have only recently come into contact with the outside

A dancer performing one of Bali's traditional dances. The Indonesian island still preserves its rich culture.

Easter Island

Some 4,000 km from South America and nearly 2,500 km from the nearest inhabited island, lies mysterious Easter Island. Scattered around the island are tall stone statues. Polynesian islanders used to light fires and worship in front of these statues. Nobody knows how the islanders managed to transport and erect the great slabs of stone.

The stone statues of Easter Island.

Warriors from the remote mountain valleys of New Guinea.

world. Probably related to the earliest settlers from Asia about 40,000 years ago, these people are very short with dark skins, frizzy hair and the features of many African peoples. Their Stone Age existence as hunter-gatherers has been preserved because of the remoteness of their valley homes, hidden in the dense tropical forests of this island.

Mixed races

When European explorers reached the Pacific islands in the 1700s, they brought trade and slave labour. Many people of the Pacific area were moved from island to island and the different races mixed. This is true in the history of many parts of the world as international trade grew and islands were settled by outsiders.

In the Indian Ocean, the Seychelles islands were used as bases by pirates until the French claimed the islands in 1754. Africans were captured and brought to work on the plantations as slaves. Later the British took control, and when slavery was abolished in 1833, they brought in Indian and Chinese labour to do the work. Today the people of the Seychelles have every shade of skin and hair imaginable, because they are a mixture of African, Indian, Chinese, Arab and European blood.

Madagascar's native people, probably of Asian origin, first came into contact with outsiders in about AD 900 when Arab traders arrived. The traders settled and brought in African slaves. Later, Europeans took control

Mutiny on the *Bounty*

One strange island story concerns the sailors on a Royal Naval ship, HMS Bounty. *After setting their captain adrift in a longboat, the remaining crew sailed away and no-one knew what became of them. Many years later, on Pitcairn Island in the South Pacific, a race of half English and half native people were found – the descendants of the* Bounty *mutineers.*

The people of Lake Titicaca in Peru have built homes on floating islands of reeds in the middle of the lake.

of the island but they did not become the dominant race in Madagascar, as they did on some other islands such as Hawaii or New Zealand.

The islands of the Caribbean have seen dramatic changes in their people. Thought to have come to the islands about 1,500 years ago, the Arawaks were relatively peaceful farmers and fishermen. However, an aggressive tribe called the Caribs arrived and ferociously attacked the Arawaks. The Caribs were fighters and fine canoeists; they killed the Arawaks and took the females as slaves. Their advance across the islands was halted by the arrival of Europeans in the late 1400s, who virtually wiped out both Arawaks and Caribs. The Europeans settled and brought in thousands of Africans to work on plantations as slaves. Later they imported labourers from Asia. Although the islands have mixed races, the Africans are dominant in the region both in numbers and culturally.

Carnival on the island of Trinidad, in the Caribbean.

The voyage of *Kontiki*

Scientists do not agree on where the Polynesians came from. The traditional view is that the Polynesians spread from Asia east across the Pacific. But Thor Heyerdahl, a Norwegian ethnologist, believed they came westward from the Americas. To prove his theory he set out from Peru, in South America, on a balsa-wood raft called Kontiki. *He reached Polynesia and proved that the winds and currents in the Pacific made it more likely that the Polynesians came from the Americas. It is also true that the native people of Polynesia and the Americas have a similar physical appearance, similar styles of religious worship and certain language similarities.*

6. ISLAND EXPLORERS AND SCIENTISTS

Polynesians

Islanders themselves usually make good explorers because, being surrounded by water, they are often experienced sailors. The original Polynesian islanders were such a people. It is thought that the Society Islands (these include Tahiti) were their base and from there they spread out north to Hawaii and south to New Zealand, where they became known as the Maoris. Their greatest travels were probably from about AD 800 onwards. There are no records of their voyages, just the evidence of how far the Polynesian race spread.

Although they were good boatbuilders and navigators, the Polynesians must have been immensely brave to set out into unknown seas without a compass or map. The Pacific Ocean is vast and they would have sailed for weeks and sometimes months, not knowing where they were heading and without seeing land. Their dugout canoes could be up to thirty metres long and usually had stabilizers called outriggers. The Polynesians were looking for new lands to colonize and on board were men, women, children, animals and plants, as well as food and water for the journey.

Polynesians use outriggers to stabilize their boats, in much the same way as their seafaring ancestors.

Christopher Columbus lands on a Caribbean island, believing it to be in the East Indies.

European explorers

People set out from all over the world to explore distant islands. The Arabs and Chinese visited many of the islands of the Indian Ocean and Southeast Asia, but it was European explorers who set forth on great sea journeys in the early fifteenth century who began to keep careful records and to chart even the most remote of the world's islands.

The first wave of European explorers set out in the hope of finding riches and trade. They wanted to find sea routes to the famed Spice Islands of the East Indies. In Europe there was a huge demand for spices (such as cloves and cinnamon) and silk cloth, and most of the trade was controlled by Arab merchants who sailed the Indian Ocean. With new, improved methods of shipbuilding, Europeans thought they could reach the islands themselves and buy direct from the islanders.

Travelling west across the Atlantic, Christopher Columbus arrived in the Caribbean Islands in 1492, as he searched for a route to the East Indies and China. He thought he had reached the East Indies, but in fact the first Europeans to get there were the Portuguese, who sailed around the African continent and across the Indian Ocean. The Portuguese found islands such as the Moluccas, Sumatra, Java and Sri Lanka and began

Spices

Spices originally only grew in South and Southeast Asia. Cinnamon was grown on the island of Sri Lanka, the Moluccas were renowned for their nutmeg and cloves, and pepper grew on most of the islands. In Europe these spices were used to preserve meat through the winter and to flavour foods, as well as being used for medicines. Spices were extremely valuable to Europeans, and the search for spices and trade with island producers was as valuable as the quest for gold.

trading in precious stones, pepper, cinnamon, cloves and sandalwood. All the European nations wanted a share of this rich East Indian trade, but the Dutch became the most successful. Within a hundred years they had taken over from the Portuguese as the main traders.

This was not a happy time for the islanders themselves, for there were constant attacks on local rulers as well as fighting among the Europeans. Just to prevent other countries from buying cloves, the Dutch laid waste to the Banda Islands, a major source of spice, killing 15,000 people and enslaving the rest. Then in 1625 they destroyed all the clove trees on the Moluccas except for the one island in their possession.

Scientific Explorers

The last major island group to be explored by Europeans were the Pacific islands, the most numerous and the most distant. Captain James Cook, an Englishman, led a series of long expeditions across the Pacific Ocean from 1768 onward. This was the beginning of a new stage in European exploration when part of the reason for the expeditions was scientific curiosity.

On board Cook's ships were biologists and naturalists who wanted to explore the remote islands in order to learn about their flora and fauna. One of the biologists on board was Joseph Banks, who brought back a large collection of butterflies, insects and plants. Captain Cook was also testing navigational equipment that proved vital for charting the exact position of some of these islands.

Captain Cook's ships arrive at Tahiti, in the Pacific Ocean, in 1773.

Left The Galapagos Islands, where Darwin studied the wildlife and formed his theory of evolution.

Islands as unique habitats were first properly explored by the two naturalists, Charles Darwin and Alfred Wallace. In 1832 Darwin set off in the ship HMS *Beagle* around the world and made a number of visits to the Galapagos Islands. He noticed that the animals of the islands resembled those of South America, but many had strange features. He wondered why they were not the same as the continental creatures and gradually formed his theory of evolution, which he published in a book, *On the Origin of the Species,* in 1859. At the same time Wallace, in Indonesia, had come to the same conclusions from his studies of wildlife on the islands in that part of the world. Although their theories are widely accepted today, in the mid-nineteenth century they were regarded as revolutionary.

Today islands are still being explored, mostly by scientists. Because remote islands are isolated, the plant and animal life is little influenced by outside forces. This makes island systems of wildlife easier to study than those of a whole continent. As a result, islands like Aldabra have become particularly important places for the study of the relationships between creatures and their habitats.

Below Little touched by outside influences, Aldabra Island, in the Indian Ocean, is an important place for scientific study today.

7. HUMAN INTERFERENCE

In many cases exploration by humans brought about much more dramatic changes in the creatures and wildlife of islands than was caused by natural evolution. And the changes occurred in a much shorter space of time.

The great Polynesian explorers were to have lasting effects in New Zealand. When the Maoris arrived there, they found a fertile land containing large animals and the flightless birds called moas. The Maoris hunted the moas for their meat and used their skins for clothing, cleared their natural forest habitat and brought rats and dogs that ate the birds' chicks and eggs. The arrival of Europeans in New Zealand, 200 years ago, added to this disruption, with the cutting down of forests and introduction of sheep to graze the land. Within a few centuries of human arrival all the moas in New Zealand were extinct except the kiwi, which is the last remaining species.

New Zealand's landscape, flora and fauna have been changed since land was cleared and non-native animals like sheep were introduced.

Huge areas of forest in Sumatra have been cut down to make way for crops.

A similar fate befell the giant tortoises of the islands of the Indian Ocean, except on Aldabra, which lay far off the main trade routes. In the early days of exploration, there was no way fresh meat could be kept on long sea journeys. The tortoises of the islands not only provided a meal for sailors who visited the islands but some could also be kept alive aboard ship so that fresh meat was available. Eventually whole tortoise populations were wiped out on these islands.

Human suffering

It is not just animals that are affected when new people arrive on an island. On many islands in the Caribbean, where Europeans landed and settled, the whole Arawak population was wiped out within fifty years. The Arawaks were killed by a combination of disease and hardships imposed by the Europeans. Today, in Kalimantan on the island of Borneo, there is massive deforestation, with the result that the native forest home of the Punan tribespeople is being destroyed and they are having difficulty surviving.

Whale bones litter the shore of South Georgia, once a whaling station in the South Atlantic.

Almost every island on earth has some species that humans have introduced. Some were brought deliberately, either to breed for food or to control pests that newcomers found on islands. Other creatures arrived by accident.

The most common animals introduced on to islands were those that settlers were used to breeding at home, such as pigs, goats, chickens and rabbits. They also brought dogs and cats, and unknowingly, rats that were living in the holds of ships. Some of these creatures remained domesticated but others became wild and caused great ecological changes such as in the Ascension Islands, where wild goats have stripped areas bare of vegetation.

There is probably no more altered habitat than New Zealand. Of some fifty-three attempts to introduce new mammals, thirty-four became established in the wild and these gradually took the place of many of New Zealand's native species. It seems that despite being well-adapted to their island, native creatures do not survive the arrival of newcomers to their habitat and the competition that they bring.

There were no native freshwater fish in New Zealand's rivers, so they were stocked with trout from Britain. The non-native rabbit became a pest when it bred in the wild and started eating crops; so weasels were introduced and let loose to control the rabbits. Deer have thrived so well that they are causing terrible damage to vegetation and are now classed as vermin. Humans did bring some useful creatures however, such as honey bees.

Introducing an animal to control another is not always a success. In the Caribbean, the mongoose was brought in to control the rat population on the islands. It did that very well, but mongooses then bred and started eating native reptiles and birds. Some native species have become extinct as a result. On Moorea in the Pacific, there once lived the small partula snail. The giant African snail was introduced to the island but it began eating the crops. So another species of meat-eating snail was brought in to eat the giant snails. Unfortunately it ate the partula too.

The dodo

The dodo was a flightless bird that lived on Mauritius. Portuguese explorers who landed there called it duodo, *which means simpleton in Portuguese. They called it this because it trusted people so much it would run to them and could be knocked on the head and killed. To sailors it was a valuable source of food on long sea journeys and they killed the last dodo in 1681. The bird is now extinct.*

8. AN ISLAND LIVING

About 10 per cent of the world's population live on islands. Some like Japan, Britain, Java and Hong Kong are densely populated. Some have become extremely powerful and successful. As well as the larger wealthy nations like Japan and Britain, there are also tiny, very rich islands like Hong Kong and Singapore. The last two both lie just off the coast of mainland Asia and have become leading world financial and trading centres. It is partly their location on important communication routes around the world that has helped them to succeed. Islands such as Sri Lanka and Hawaii have also benefitted from lying at strategic points on air and sea routes.

Above Singapore is a Chinese-run modern island city in Southeast Asia.

One of the advantages islanders have for making a living is a good supply of fish in the sea around them. As well as providing food, surplus fish can be sold and traded for other goods that may be in short supply on an island

Left Icelandic fishing boats. Many islanders depend on fishing for part of their livelihood.

The people of South Uist in the Outer Hebridges survive by stock raising, crofting and fishing.

with limited resources. The settlement of the bleak Canadian island of Newfoundland was entirely based on the rich catches of cod fish in the seas offshore. It was the same for settlers on remote islands such as the Falkland Islands and South Georgia in the South Atlantic. The large population on the islands of the Pacific could not survive without fishing and in the Indian Ocean for example, the Seychelles islanders catch tuna and export it tinned.

Water shortages

Like many small islands around the world, Key West in Florida, an island that is home to the most southerly city in the United States, suffers from a lack of fresh water. Desalination plants have been built to convert sea water to drinking water. Other poorer island communities around the world cannot afford this expensive way of ensuring there is enough water.

The Florida Keys, islands off the southern tip of Florida, are linked by road bridges across the sea.

Plantation crops

The more remote the island, the more people are likely to exist at a subsistence level, growing food crops, perhaps raising a few animals and fishing. But on many of the islands that were colonized by Europeans, plantation crops such as sugar, coffee and bananas were introduced. These cash crops were and still are sold for export world-wide. In the early days the majority of islanders would not have have reaped the benefits of the plantation crops; the profits would have gone to the rich European plantation owners, some of whom may not have even lived on the island.

The economies of many islands were often dominated by plantation crops. The problem was then that if demand fell or, as sometimes happened, a cyclone destroyed the harvest, the islanders would have no other way to make

Tea-picking on a tea plantation in Sri Lanka.

Loading sugar cane in Haiti.

a living. Cuba in the Caribbean and Mauritius in the Indian Ocean were two such islands that depended on growing sugar cane. Now the islanders have tried to introduce other crops and industries to help balance their economies.

The island of Sri Lanka off the south coast of India grows tea, rubber and coconuts. It has to import some of its basic foodstuff, rice, and still depends on the unreliable world market to make money out of its cash crops. In Papua New Guinea, little space has been left for the crops people need to grow to eat because so many cash crops such as coffee, cocoa and coconut are being grown.

Some islands have rich natural resources. But seldom have the islanders themselves been allowed to control and use these resources; more often they have been exploited by larger and wealthier countries. Borneo is thickly covered in tropical forest, but the island is being stripped of its wood, mainly by the Indonesian government, to sell to meet Japanese demand for hardwoods. In Papua New Guinea there is a huge copper mine, but it is not owned by the local people and they have had little more than jobs as poorly paid workers in the mine.

Yachts in the harbour at Antigua, in the Caribbean, where tourism is a major source of income.

Tourism

One of the most important sources of income for islands today is tourism. It is not only the beautiful islands with sandy beaches and palm trees that attract tourists. There is also a growing interest in visiting islands with curious natural features such as the wildlife of the Galapagos and the splendid volcanic scenery of Iceland. Cheaper air travel is one of the main factors that has made it possible for people to reach some of the world's remote and interesting islands.

Some of the main island groups that rely on tourism are the Caribbean Islands, the islands of the Seychelles and Maldives in the Indian Ocean, the Pacific islands, the Canary Islands in the Atlantic and the many islands of the Mediterranean Sea.

Tourism has brought jobs (such as in hotels, banks, shops and restaurants) and money to many of these islands, so that local standards of living have risen. In the Seychelles local people have given up farming and fishing and about 20 per cent of the workforce is now employed in better paid jobs in tourism.

However tourism does have its drawbacks. Some islands have become so dependent on tourism for their economy that if tourists are put off from coming to the island for any reason (perhaps political trouble or a natural disaster like a cyclone), the economy slumps. If this happens, islanders seldom want to return to farming and fishing.

It is not always the local people who benefit from tourism. The greater part of the price of a package holiday goes to the foreign operator; in some cases hotels are owned by a foreign travel company and so the profits go back to the company's country of origin. Although tourism means that new buildings, roads and airports have to be built, which helps local businesses, it also means land is lost from agricultural use. In addition, many new building developments are very ugly. On the other hand, historic buildings that might otherwise fall into disrepair may be preserved for tourists to visit.

Tourism also brings a change in lifestyle and traditions for islanders. The island of Bali in Indonesia has a particularly rich cultural history and is one reason why tourists visit the island. The islanders put on special performances of their dances for tourists and sell handicrafts. Some say tourism has helped preserve the customs, ceremonies and arts of the island; others say the Balinese culture is dead because it is now just a money-making business.

Local boys make money selling towels to tourists on the island of Ko Samui, Thailand.

9. CONTROL AND OWNERSHIP

An airforce base on Ascension Island. The island serves as an important communications base in the middle of the Atlantic Ocean.

As well as Australia which, because of its vast size, is usually considered to be a continent not an island, there are nine large nations in the world that consist entirely of islands. These are Indonesia, Japan, Philippines, Australia, New Zealand, Sri Lanka, Madagascar, Britain, Iceland and Cuba. The largest in population and in area is Indonesia: in total it is almost the size of Greenland but it is made up of 13,677 islands. Some 6,000 of these are inhabited and the population is the fifth biggest in the world. Greenland, the world's largest island, belongs to Denmark though it is largely self-governing and may soon achieve independence. There are also hundreds of small islands, especially in the Caribbean and the Pacific, that are countries in their own right.

Throughout history islands have often been fought over. Who now inhabits and controls an island depends very much on that island's history. Almost all islands were at one time colonies of European countries, although many have now achieved their independence and are self-governing.

Above Jamaica was first occupied by Arawaks and then the Spanish. It later fell to the British, who imported Africans to work the plantations. The island became independent from Britain in 1962.

Left A Greek sign reminds people of the struggle with Turkey for control of the island of Cyprus, in the eastern Mediterranean.

The Falkland Islands still belong to Britain, even though they lie thousands of kilometres away in the South Atlantic.

The Caribbean was one area in particular where islands frequently changed hands. After Columbus landed in 1492, Spanish settlers arrived and started colonies on the islands. They were followed by British, French, Dutch, Danes and other Europeans, who grabbed islands as their footholds in the New World. Wars in Europe over the following centuries, and rivalry in the Caribbean, meant continual battles for control of the islands. Things settled down in the 1800s and during the 1900s many islands gained independence from their colonial rulers. There are now 24 separate states in the Caribbean.

Some islands have a split ownership, for example Hispaniola in the Caribbean. The eastern part is the French-speaking Republic of Haiti and the western half is the Spanish-speaking Dominican Republic that has close links with the United States. New Guinea in Southeast Asia is divided between Irian Jaya, belonging to Indonesia, and the independent state of Papua New Guinea. In the same area is Borneo, an island divided in three: the independent state of Brunei, Sabah and Sarawak that belong to Malaysia, and

Kalimantan that is owned by Indonesia. In the Mediterranean is the tiny island of Cyprus. It has been partitioned since 1974, between Turkey and Greece, because neither nation can agree who is the rightful owner.

Often islands are occupied not for the resources such as minerals or timber that they can provide but because they add to a country's territory. Or they may be in a strategic position or a useful communications base. For example Hong Kong and the Falkland Islands have always been useful bases in far-flung parts of the world for Britain, as have Guam (in the north Pacific) and Hawaii for the United States. France keeps hold of the islands in Tahiti-Polynesia and uses outlying ones for nuclear tests.

Supposedly independent islands often rely on powerful foreign countries for help or aid. The Philippines have for a long time been supported financially by the United States and in return the Americans, until recently, sited huge military bases there. Payments are made to the Seychelles government for the use of land for a United States satellite tracking station and Cuba was sustained for many years by the former Soviet Union, which helped the island economically and kept military forces on the island.

After 155 years of British ownership, the busy island of Hong Kong will be returned to China in 1997.

10. ENVIRONMENT AND SURVIVAL

Like all habitats, islands are under environmental threat from misuse or overuse by humans. The small size of most islands means that harm can occur quickly and can have devastating effects if the situation is not controlled.

On less developed islands tourism can bring additional pressures where there is already a shortage of resources such as gas, coal or electric power. Water is a particular problem for many islands that are popular with tourists. One solution is to build desalination plants to use the water from the sea, but these are very expensive for developing countries. Tourism can also bring pollution because the sewage systems are unable to cope with the additional number of people; uncleared rubbish left by tourists often accumulates on beaches.

Rubbish has to be cleared regularly from the crowded tourist beaches on the Mediterranean island of Majorca.

Soil erosion is particularly bad on Madagascar, where trees have been cleared for farming.

Tourism can have good and bad effects on the flora and fauna of islands. Sometimes tourism encourages the preservation of rare creatures in wildlife sanctuaries for visitors to see, such as the lemurs of Madagascar. In the Florida Keys, islands lying off the coast of the United States near Miami, there are many state and country parks where coral and marine animals are protected. On the other hand, tourism results in harm when coral or large sea shells are bought as souvenirs – around many islands the sea has been stripped of large shellfish to meet the demand. Coral is in danger all over the world because it is particularly sensitive to changes in its environment and the slightest pollution in the water is enough to stop it growing.

Plantation agriculture has caused environmental damage because it needs almost all the natural vegetation to be removed before planting begins. Mauritius has virtually no natural woodland left; Hawaii is suffering from soil erosion due to clearing the land of its natural cover and ploughing for pineapple planting. Overgrazing by livestock has caused soil erosion on islands such as Madagascar, where vast areas of forest were cleared in order to make pasture for cattle introduced from Africa.

Other islands where there has been serious deforestation are Sumatra and Borneo. It is thought that overall in Indonesia, more than one-third of all the

Atomic testing

France has carried out more than eighty nuclear tests on the island of Moruroa in the South Pacific. It was hoped that the island was so remote that the tests would do no harm. Since testing nuclear weapons by exploding them underground, the island has begun to sink and is now 1.5 m lower than it used to be. The whole island has been weakened and radioactivity is leaking into the sea through a crack in the atoll. Tropical storms make matters worse. The radioactivity concentrates in seafood, the main diet of Pacific people and can produce serious illnesses.

An underwater explosion during atomic tests in Bikini lagoon, in the Pacific Ocean.

tropical forest has been chopped down since 1950. The effects can be devastating for the environment: wildlife dies when it loses its natural habitat, soil erodes without vegetation cover, flooding occurs and there may even be local climate changes. Problems like these are often recognized on under-developed islands but they are not being tackled, partly through a short-term desire to make money.

The seas around islands are under pressure too. Overfishing is a worldwide problem that not only harms the natural balance of life in the sea but also affects the livelihood and survival of island people. Nowadays fishing island-nations like Iceland, Britain and Japan have huge factory trawlers that catch millions of tonnes of fish a year. In the North Atlantic stocks of fish have fallen dramatically in the last few years and there are now controls on the amount of fish that are allowed to be caught.

Oil-covered birds were casualities after a tanker went aground off the Shetland Islands, in the North Atlantic, in 1993.

The size of modern ships and their potentially dangerous cargoes are a worry for islanders. In the Shetland Islands, off the coast of Scotland, there was a near-disaster when an oil tanker went aground in rough seas in 1993. The Pacific Islands with their hidden coral reefs have seen thousands of shipwrecks over the years. For many years at Key West in Florida, the most important industry was salvaging cargo from ships that had foundered on nearby reefs.

In some ways, developed, industrialized islands such as Britain and Japan, face the most severe environmental problems. They are the sort that are found in any overcrowded part of our planet, in particular air, water and soil pollution from industry and agriculture. The difference is that the effects of these problems are magnified because Britain and Japan are islands and the environment has tight limits on how much it can take.

However, such rich nations can afford to take the costly measures needed to control environmental problems. These nations are also spending time and money looking at solutions in order that humans can live safely in increasingly crowded and limited places such as islands. Studying islands and the relationships between the creatures, plants and environment may show us a way forward for our whole planet, which is, after all, one huge island.

A peaceful island far from land: Moorea, in the South Pacific, is a favourite haven for tourists.

Glossary

Atoll A circle of coral reefs.
Cash crops Crops gown for sale, rather than for living on.
Colony Land settled and controlled by a powerful country.
Continent A very large land mass.
Deforestation The cutting down and clearance of trees from land.
Delta A flat, fan-shaped area of land near the mouth of a river, where the main stream splits into many channels at the end of its course.
Desalination The process of removing salt, especially from seawater.
Dialects Local accents
Ecologies The ways in which living things relate to each other and their environment.
Erosion The process of soil being washed or blown away.
Ethnologist Someone who studies the behaviour of animals.
Evolve When the characteristics of animals and plants gradually change over a long period. This gradual change is called evolution.
Extinct When a species of living thing dies out completely.
Fauna All the animal life of a particular place or time.
Flora All the plant life of a particular place or time.
Fossils The remains, traces or impressions of animals found preserved in rock.
Immigration Moving from one country to live permanently in another.
Inbred Bred from closely related parents over a period of time.
Islets Very small islands.
Lagoon A stretch of water cut off from the open sea by coral reefs or sand bars.
Latitude The distance of a place north or south of the equator.
Lava Molten rock from inside the earth that pours out of a volcano.
Lichen Tiny, fungus-like plants that grow on the surface of rocks and tree-trunks.
New World An old term for North and South America.
Package holiday A holiday arranged by a holiday tour company as an all-inclusive deal.
Plantations Estates, especially in tropical countries, where cash crops such as rubber, tea and cotton. are grown on a large scale, for sale.
Pollution Poisonous substances released into air, water and land that may upset the natural balance of the environment.
Predator A creature that hunts and kills another.
Radioactivity The giving out of energy in the form of dangerous rays.
Reef A ridge of rocks, sand or coral, the top of which lies close to the surface of the sea.
Sagas Ancient stories, often about battles, warrior gods and heroes.
Sanctuaries Places of refuge or safety.
Sediment Fine particles of mud and earth that are deposited by water.
Sewage Waste material and liquid from factories and houses, carried in drains and sewers.
Soviet Union A former group of republics in Eastern Europe and Northern Asia controlled from Moscow. The Soviet Union broke up into separate states in 1991.
Species A group of plants or creatures with similar characteristics.
Stone Age The earliest period in human history.
Strait A narrow channel of sea linking two larger areas of sea.
Strategic To be in a well-placed position.
Stratosphere An upper region or layer of the atmosphere, high above the earth's surface.
Subsistence level A level of income that will only buy the basic needs of life, such as food.
Vermin Troublesome or destructive animals such as mice or rats.

Books to Read and Further Information

Books for younger readers

Life in the Islands by Rosanne Hooper (Two-Can, 1992)
Story of the Earth: Island by Lionel Bender (Franklin Watts, 1989)
Coral Reefs by Jenny Wood (Franklin Watts, 1991)
Vanishing Habitats by Noel Simon (Franklin Watts, 1987)

Books for older readers

Island Life by Marion Steinmann (Time-Life Films, 1978)
The Living Planet by David Attenborough (Collins, 1984)

Other sources of information are travel and geography books about specific groups of islands, for example, the Caribbean Islands, Greek Islands, Pacific Islands, Canary Islands, Indonesia, Philippines and Seychelles. It is also worth looking at books about volcanoes and coral for background information. The BBC produce two interesting videos which are well worth watching: *The Living Planet* and *Life on Earth: A Natural History*, both by David Attenborough.

For further information about animals and their habitats that are under threat, contact the following environmental organisations:

Friends of the Earth, 26-28 Underwood Street, London N1 7JQ.
Worldwide Fund for Nature, Panda House, Weyside Park, Godalming, Surrey GU7 1XR.
Greenpeace, 30-31 Islington Green, London N1 8XE.

These organizations all campaign to protect wildlife and habitats throughout the world.

Picture acknowledgements
Britstock-IFA cover, Eye Ubiquitous/J. Davis 5(top), 5(lower), /J. Davis 8, 12, 20(top), /D. Brannigan 21(top), /E. Enstone 23(lower), /J. Davis 24, 29, /D. Cumming 32(top), /B. Adams 36, /J. Hulme 37, /J. Davis 39; Frank Lane Picture Library 9, /W.Wisniewski 18, 27(lower), /S.Jonasson 32(lower); NHPA /P. Johnson 14(top), 17(lower), /K.Switak 19(top), /H.Palo 27(top), /D. Woodfall 33(top), /P. Johnson 38; Still Pictures /D. Harms 7, /Y. Lefevre 15, /T. Thomas 16, /Deulefeu 17(top), /R. Seitre 19(lower), /Michel Gunther 43, /Mark Edwards 44(lower); Topham Picture Library 11, 13(top), 14(lower, both), 20 lower), 21(lower), 22, 23(top), 25, 30, 31, 34, 35, 42, 44(top); Julia Waterlow 39(lower), 41; Wayland Picture Library title page, 13(lower), 26, 28, /D. Cumming 39(top); Zefa 10, 33(lower), 40, 45. Maps and diagrams on pages 4, 6, 10 Peter Bull, and 15 John Yates.

INDEX

Numbers in **bold** refer to photographs